Tasmanian Devils

By Christy Steele

Steadwell Books

Raintree

A Division of Reed Elsevier, Inc.

Chicago, Illinois

www.raintreelibrary.com

ANIMALS OF THE RAIN FOREST

For information, address the publisher:
Raintree, 100 N. LaSalle, Suite 1200, Chicago, IL 60602

Library of Congress Cataloging-in-Publication Data
Steele, Christy.
   Tasmanian devils / Christy Steele.
      v. cm. -- (Animals of the rain forest)
Includes bibliographical references (p.   ).
Contents: Range map for Tasmanian devils -- Tasmanian devils in the rain forest -- What Tasmanian devils eat -- A Tasmanian devil's life cycle -- The future of Tasmanian devils.
   ISBN 0-7398-6840-3 (lib. bdg. : hardcover)
   1. Tasmanian devil--Juvenile literature. [1. Tasmanian devil.] I.
Title. II. Series.
   QL737.M33 S74 203
   599.2'7--dc21
                              2002015213
Printed and bound in the United States of America

Produced by Compass Books

Photo Acknowledgments
Tom Stack/Dave Watts, cover, 14, 18, 21; Root Resources/Mary and Lloyd McCarthy, 1, 24, 28-29, Tom Stack/Chip Isenhart, 6; Tom Stack/Gary Milburn, 8, 11, 12; Visuals Unlimited/Tom J, Ulrich, 22; Unicorn/Margo Moss, 26.

Content Consultants
Nicholas Klomp
Head of School
School of Environmental and Information Sciences
Charles Sturt University, Albury, NSW

Mark Rosenthal
Abra Prentice Wilkin Curator of Large Mammals
Lincoln Park Zoo, Chicago, Illinois

This book supports the National Science Standards.

Some words are shown in bold, **like this**. You can find out what they mean by looking in the Glossary.

Contents

Range Map of Tasmanian Devils

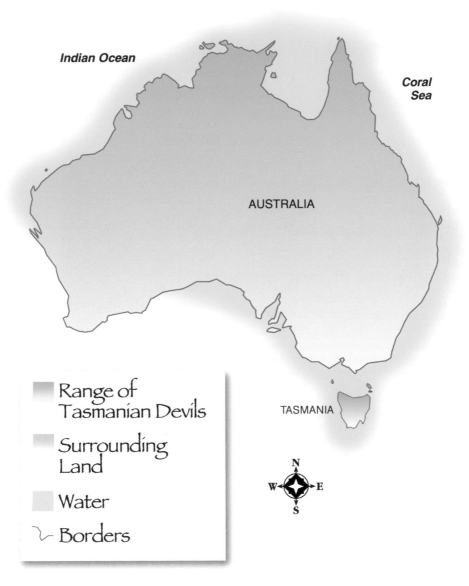

Indian Ocean

Coral Sea

AUSTRALIA

TASMANIA

Range of Tasmanian Devils

Surrounding Land

Water

Borders

N
W E
S

A Quick Look at Tasmanian Devils

What do Tasmanian devils look like?

Tasmanian devils are small animals with bear-like faces. They have short black fur with white markings. Their large mouths are full of many sharp teeth.

Where do Tasmanian devils live?

Tasmanian devils live only in Tasmania. This is an island state of Australia.

What do Tasmanian devils eat?

Tasmanian devils eat only meat. They will eat almost any kind of meat that they can find.

Tasmanian devils have many sharp teeth.

Tasmanian Devils in the Rain Forest

Tasmanian devils may have a scary name, but they are not ghosts or monsters. Usually, these small animals stay away from people. They will not attack people unless they sense danger.

The scientific name for Tasmanian devils is *Sarcophilus harrisii*. In Latin, *sarcophilus* means meat lover. *Harrisii* is for a man named Edwin Harris. He was one of the first scientists to study and write about Tasmanian devils.

Tasmanian devils are named after Tasmania, an island state of Australia. This is the only place where they live in the wild. European settlers named them devils because these toothy animals make loud noises at night.

 This Tasmanian devil is resting in its burrow.

Where do Tasmanian devils live?

Tasmanian devils live in many places throughout the island of Tasmania. They have **adapted** to the different **habitats** on the island, including mountains and rain forests. Adapted means a plant or animal has changed to fit the area, or habitat, where it lives.

In the rain forest, Tasmanian devils make their homes among the trees. During the day, they stay inside their dens. Dens may be hollow logs, caves, or burrows that other animals have left. A burrow is a hole that an animal has dug to live inside. If Tasmanian devils cannot find a den, they will often sleep in thick bushes.

Dens and bushes keep Tasmanian devils safe while they sleep. Otherwise, **predators** could catch them. A predator is an animal that hunts another animal for food. Owls, hawks, quolls, or native cats will eat any Tasmanian devils they catch.

Tasmanian devils often live near water. They swim very well and sit in the water to cool themselves when it is hot.

Tasmanian devils spend most of their time on the ground. Even so, young Tasmanian devils are excellent climbers. They can grip the tree with their claws and front paws. Then they push themselves up the tree with the large footpads on their back paws. They can even race up trees to escape predators. Some older Tasmanian devils can still climb, but not as quickly.

What do Tasmanian devils look like?

At about 1 foot (30 centimeters) tall, Tasmanian devils are the size of a small dog. They have bushy tails and thick powerful bodies up to 2 feet (60 centimeters) long. Males are a little larger and heavier than females. Males weigh up to 25 pounds (11 kilograms), while most females weigh up to 18 pounds (8 kilograms).

A Tasmanian devil's face looks like that of a small bear. Its large head may weigh up to one-fourth of its body weight. Powerful jaw muscles attach to ridges on its thick skull. The jaws open extra wide and are strong enough to crush bone.

Tasmanian devils have short, black fur with white markings on the chest and backside. This coloring helps **camouflage** them in the rain forest. Camouflage is colors, shapes, and patterns that help an animal blend in with its background. The black color helps them blend in with the dark trees and bushes. The white stripe breaks up the outline of their bodies.

Tasmanian devils have short legs. There are five toes on their front paws and four toes on their back paws. A sharp claw grows on each toe.

The coloring of this Tasmanian devil helps it blend in with its surroundings.

 This Tasmanian devil is yawning.

How do Tasmanian devils act?

Tasmanian devils have a home range of about 50 acres (20 hectares). They live in this range and travel around it to look for food. The size of the range depends on the amount of food and how many Tasmanian devils are in one place. Even though they may share parts of their home range with other Tasmanian devils, they spend most of their time alone.

Tasmanian devils gather together only to mate or feed on a large animal. When this happens, the dominant animals are the first to eat. Dominant means most powerful. To show dominance, Tasmanian devils use body language, sounds, and smells.

Body language is one way to communicate. If a Tasmanian devil yawns, it is upset or scared. Even the way it holds its tail sends a message to other Tasmanian devils. For example, if a Tasmanian devil holds its tail straight up, it is angry and about to fight.

Tasmanian devils are some of the noisiest animals in the world. Each sound means something to other Tasmanian devils. They growl, cough, snort, and screech. A loud sneeze calls another Tasmanian devil to fight.

Smells also send messages to other Tasmanian devils. When calm, the animal does not have a strong scent. If the Tasmanian devil becomes excited, its body releases a powerful, stinky scent.

This Tasmanian devil is eating meat that it has found.

What Tasmanian Devils Eat

Tasmanian devils are **carnivores**. A carnivore eats only meat. Each day, Tasmanian devils need to eat about 15 percent of their body weight to stay alive. To do this, they will eat almost any kind of meat they can find.

Younger Tasmanian devils eat small animals that are easier to catch. Insects, frogs, fish, snakes, rabbits, and birds make a tasty meal. Larger Tasmanian devils eat both small and large animals, including wombats, wallabies, platypus, lambs, and sheep.

All Tasmanian devils eat carrion, or flesh from dead animals. Sometimes these animals have died from old age or sickness. By eating carrion, Tasmanian devils help keep the rain forest clean of rotting bodies.

 These Tasmanian devils have gathered together to eat.

Hunting and eating

Tasmanian devils hunt at night. Their black fur makes it hard for other animals to see them. Tasmanian devils can see movement well, and their excellent sense of smell helps them find food.

To hunt, Tasmanian devils move slowly along animal trails. They usually look for old, sick, or young prey because these animals are slower and easier to catch. They sniff the air to find their prey's scent. Prey is an animal that is hunted and eaten as food. Once they find prey, they can run up to 8 miles (13 kilometers) per hour for short distances. If they catch prey, they use their powerful jaws to bite it in the neck.

Tasmanian devils are known for the way they eat. Large groups of them gather together and noisily rip apart prey. They growl at each other and may even fight if there is not enough food.

They have teeth suited for tearing prey and crushing bones. Because of this, they can eat all of their prey, including fur and bones. A hungry Tasmanian devil can eat up to 40 percent of its body weight at a time. A huge meal like this would last it up to 3 days.

Tasmanian devils can store fat in their tails. A fat tail is a sign of a healthy animal. If the tail is too thin, this means the Tasmanian devil is sick or needs food.

This female has found a hidden den where she has given birth to young.

A Tasmanian Devil's Life Cycle

Most Tasmanian devils begin mating when they are about two years old. Their mating season is in the spring, usually in March.

Most fighting happens during mating season. Males fight with other males to win the right to mate with females.

After mating, females fight the males to drive them away. Then, females search for hidden dens where they can give birth.

Like kangaroos, Tasmanian devils are **marsupials**. All female marsupials have pouches. Their young stay in the mother's pouch for several months after they are born.

Young Tasmanian devils

In April, the female gives birth to up to 30 young Tasmanian devils. These young are not fully developed. They need more time to grow before they can live on their own. As soon as they are born, they try to crawl into their mother's pouch. But the mother has room for only four young in her pouch. The ones that do not make it into the pouch die. In the wild, only the strongest animals survive.

Once her young are inside, the female closes her pouch. This keeps them safe from **predators**. Inside the pouch, the young Tasmanian devils drink their mother's milk and continue to grow.

After about four months, the young leave their mother's pouch. For several months, they continue **nursing**. They stay in a den while their mother leaves to eat. By the time they are eight months old, they stop drinking their mother's milk. They then leave the den to find their own home ranges.

Wild Tasmanian devils live about five years. In zoos, they may live up to eight years.

This young Tasmanian devil is resting with its mother in a den.

▲ The long whiskers on this Tasmanian devil help it move around the rain forest.

A Tasmanian devil's day

During the day, Tasmanian devils sleep mainly in their dens. They sometimes leave their dens for short periods of time. If it is cool, they may warm themselves by lying in the sun. If it is hot, they spend part of the day cooling themselves by swimming or splashing in water.

Tasmanian devils have 42 teeth that continue to grow throughout their lives. Some people say that a Tasmanian devil's bite is as powerful as an alligator's bite.

Tasmanian devils are **nocturnal**. This means they are active at night. When it is dark, they leave their dens to eat. They travel around their home range to search for food. Scientists have tracked the movement of Tasmanian devils. They found that the animals may travel up to 12.5 miles (20 kilometers) a night. However, if they find food right away, they will not travel farther.

Tasmanian devils are best suited to traveling at night. They have many long whiskers all over their head. They use their whiskers to help them move through the dark rain forest. The whiskers sense objects that the animals need to move around. Tasmanian devils also know that if the whiskers can fit through a small space, the rest of their body will, too.

The Tasmanian devil is the symbol of Tasmania's National Park Service.

The Future of Tasmanian Devils

Scientists think Tasmanian devils once lived all throughout Australia. They have found Tasmanian devil **fossils** there. Fossils are the remains or imprints, such as footprints or bones, left by animals that lived long ago.

What happened to Australia's Tasmanian devils? Scientists believe wild dogs and people may have overhunted them. Over time, the Australian Tasmanian devils died out.

Early settlers in Tasmania wanted to kill all Tasmanian devils, too. Farmers thought the animals would kill their livestock. In 1830 a land company began paying money to hunters for each Tasmanian devil they killed.

A scientist is helping this sick Tasmanian devil.

What will happen to Tasmanian devils?

For about 100 years, people overhunted Tasmanian devils. By the 1930s, they were in danger of becoming **extinct**. Extinct means all of an animal or plant has died out in the wild. In 1941 new laws made it illegal to hunt

Sometimes a Tasmanian devil's ears can look red. When a devil becomes excited or stressed, blood rushes to its thin ears. This makes the ears red.

Tasmanian devils. With these new laws, the Tasmanian devil population slowly began to grow.

Today, Tasmanian devils are no longer in danger of becoming extinct. But people can still harm them. One of the main problems is loss of **habitat**. People tear down the Tasmanian devil's rain forest home to build houses and farms. This makes it hard for Tasmanian devils to find dens and food. Some farmers may kill Tasmanian devils. Others die when they cross roads and are run over by cars.

Even with these problems, the Tasmanian devil population is healthy. Scientists think there may be as many as 150,000 living in Tasmania. If people keep working to save their habitat, Tasmanian devils will survive long into the future.

white markings
see page 10

Glossary

adapted—when a plant or animal has changed to fit the area where it lives

camouflage—colors, shapes, and patterns that help an animal blend in with its background

carnivores—animals that eat only meat

extinct—when all of one kind of animal has died out

fossil—the remains or traces of an animal or plant from thousands of years ago

habitats—places where an animal or plant usually lives

marsupial—a group of animals in which the female carries its young in a pouch

nocturnal—active at night

nursing—when a mother feeds her young milk made inside her body

predators—animals that hunt other animals for food

Internet Sites

DPIWE—Tasmanian Devil
http://www.dpiwe.tas.gov.au/inter.nsf/WebPages/BHAN-5358KH?open

Tour of Tasmania—Tasmanian Devil
http://www.tased.edu.au/tot/fauna/devil.html

Useful Address

Tasmanian Devil Park
RSD 993
Taranna, Tasmania
Australia 7180

Books to Read

Preller, James. *In Search of the Real Tasmanian Devil*. New York: Scholastic, 1996.

Stone, Lynn M. *Tasmanian Devil*. Vero Beach, FL: Rourke, 1990.

Index